THE
BATTLE
FOR
PARADISE

THE BATTLE FOR PARADISE

Puerto Rico Takes On the Disaster Capitalists

NAOMI KLEIN

Haymarket Books
Chicago, Illinois

Published in 2018 by
Haymarket Books
P.O. Box 180165
Chicago, IL 60618
773-583-7884
www.haymarketbooks.org
info@haymarketbooks.org

ISBN: 978-1-60846-586-6

Trade distribution:
In the US, Consortium Book Sales and Distribution,
www.cbsd.com
In Canada, Publishers Group Canada, www.pgcbooks.ca
In the UK, Turnaround Publisher Services,
www.turnaround-uk.com
All other countries, Ingram Publisher Services International, IPS_
Intlsales@ingramcontent.com

This book was published with the generous support
of Lannan Foundation and Wallace Action Fund.

Printed in Canada by union labor.
Cover design by Rachel Cohen.

Library of Congress Cataloging-in-Publication data is available.

10 9 8 7 6 5 4 3 2 1

CONTENTS

All royalties from the sale of this book in English and Spanish go directly to JunteGente, a gathering of Puerto Rican organizations resisting disaster capitalism and advancing a fair and healthy recovery for their island. For more information, visit juntegente.org.

FOREWORD

Weeks after the passing of Hurricane Maria in Puerto Rico, members of PAReS—a collective of professors created to defend public education during the 2017 University of Puerto Rico student strike—met to discuss how to confront the devastation that the country and our university faced. What concerned us was not only the enormous physical damage caused by the storm but also the intensification of neoliberal policies to come.

We knew that the real disaster was not the hurricane but the terrible vulnerability imposed by Puerto Rico's colonial relationship to the United States, as well as the forced privatization of health and other services; massive layoffs; huge numbers of school closures; reductions in social rights and in investments for collective well-being; abandonment of social and physical infrastructure; and high levels of government corruption and ineptitude.

This vulnerability was aggravated by Washington's imposition of the Financial Oversight and Management Board, an unelected body pushing for the privatization of electricity and schools, increased costs of basic services, massive cuts in public education, pensions, vacation time, and other rights—all in order to pay bondholders a $73 billion debt that was patently unpayable, illegal, and illegitimate. The net result was to leave the majority of people in Puerto Rico without a hopeful future, and that was all before Hurricane Maria hit our shores.

PAReS decided to create a series of public forums on disasters, hoping to generate public debate and encourage new kinds of collective thinking about resistance and alternatives. We invited Naomi Klein as our first speaker, to talk about her work focusing on the application of a "shock doctrine" in various post-disaster settings. Our goal was to highlight how disaster capitalism was being applied in Puerto Rico, to promote equitable and ecological alternatives to these policies, and to strengthen the project of public education as a common good. We also wanted to denounce the exploitation of Hurricane Maria to promote widely rejected neoliberal policies that undermine our country's well-being, especially that of our most vulnerable inhabitants. These policies will limit access to basic rights such as water, electricity, and housing, and will destroy our environment, health, and democracy, as well as our quality of life and economic stability. And all the while, they will increase the transfer of wealth to the already rich.

In solidarity, Naomi accepted our invitation and spent an intense week with us in January 2018. Our time together included a forum on disaster capitalism at the University of Puerto Rico, Río Piedras, which was attended by more than 1,500 people and was widely covered in the press. We also took multiple trips across the island to research the topics of debt and privatization, energy sovereignty, and food sovereignty. The week finished with a full-day gathering of more than 60 organizations resisting disaster capitalism. These organizations have continued to meet, giving rise to the creation of the JunteGente network, with the aim of uniting different struggles for the future of Puerto Rico. Naomi's visit, as well as the presence of other groups featured in this book, helped to develop ongoing discussions on how organized civil society can build a "counter-shock" strategy able to resist disaster capitalism and promote alternatives to neoliberalism on a national scale.

A product of these intense investigations and conversations, this book clearly shows the historical juncture at which Puerto Rico finds itself. Interspersing stories of the super-rich who seek to buy our country for a bargain with reports from grass-roots struggles over agro-ecology, renewable energy, and public education, Klein acutely and captivatingly exposes the essence of the battle that is being waged between these opposing visions. On one side lies the utopia (for us, a dystopia) of Puerto Rico as a resort for the wealthy. On the other, a utopian vision of a Puerto Rico that is equitable, democratic, and sustainable for all. In addition, Klein addresses the historical complexities of this moment, linking current struggles to long-standing processes of colonialism and neoliberalism. The book is thus a necessary read for anyone who wishes to understand the ongoing crisis in Puerto Rico and to grasp what is at stake, which is nothing less than the survival of the people of our beautiful Caribbean archipelago.

Federico Cintrón Moscoso
Gustavo García López
Mariolga Reyes Cruz
Juan Carlos Rivera Ramos
Bernat Tort Ortiz
Professors Self-Assembled in Solidarity
    Resistance (PAReS)

April 2018

THE
BATTLE
FOR
PARADISE

A SOLAR OASIS

Like everywhere else in Puerto Rico, the small mountain city of Adjuntas was plunged into total darkness by Hurricane Maria. When residents left their homes to take stock of the damage, they found themselves not only without power and water, but also totally cut off from the rest of the island. Every single road was blocked, either by mounds of mud washed down from the surrounding peaks, or by fallen trees and branches. Yet amid this devastation, there was one bright spot.

Just off the main square, a large, pink colonial-style house had light shining through every window. It glowed like a beacon in the terrifying darkness.

The pink house was Casa Pueblo, a community and ecology center with deep roots in this part of the island. Twenty years ago, its founders, a family of scientists and engineers, installed solar panels on the

1

center's roof, a move that seemed rather hippy-dippy at the time. Somehow, those panels (upgraded over the years) managed to survive Maria's hurricane-force winds and falling debris. Which meant that in a sea of post-storm darkness, Casa Pueblo had the only sustained power for miles around.

And like moths to a flame, people from all over the hills of Adjuntas made their way to the warm and welcoming light.

Already a community hub before the storm, the pink house rapidly transformed into a nerve center for self-organized relief efforts. It would be weeks before the Federal Emergency Management Agency or any other agency would arrive with significant aid, so people flocked to Casa Pueblo to collect food, water, tarps, and chainsaws—and draw on its priceless power supply to charge up their electronics. Most critically, Casa Pueblo became a kind of makeshift field hospital, its airy rooms crowded with elderly people who needed to plug in oxygen machines.

Thanks also to those solar panels, Casa Pueblo's radio station was able to continue broadcasting, making it the community's sole source of information when downed power lines and cell towers had knocked out everything else. Twenty years after those panels were first installed, rooftop solar power didn't look frivolous at all—in fact, it looked like the best hope for survival in a future sure to bring more Maria-sized weather shocks.

Visiting Casa Pueblo on a recent trip to the island was something of a vertiginous experience—a bit like stepping through a portal into another world, a parallel Puerto Rico where everything worked and the mood brimmed with optimism.

It was particularly jarring because I had spent much of the day on the heavily industrialized southern coast, talking with people suffering some of the cruellest impacts of Hurricane Maria. Not only had their low-lying neighborhoods been inundated, but they also feared the storm had stirred up toxic materials from nearby fossil fuel-burning power plants and agricultural testing sites they could not hope to assess. Compounding these risks—and despite living adjacent to two of the island's largest electricity plants—many still were living in the dark.

The situation had felt unremittingly bleak, made worse by the stifling heat. But after driving up into the mountains and arriving at Casa Pueblo, the mood shifted instantly. Wide open doors welcomed us, as well as freshly brewed organic coffee from the center's own community-managed plantation. Overhead, an air-clearing downpour drummed down on those precious solar panels.

Arturo Massol-Deyá, a bearded biologist and

president of Casa Pueblo's board of directors, took me on a brief tour of the facility: the radio station, a solar-powered cinema opened since the storm, a butterfly garden, a store selling local crafts and their wildly popular brand of coffee. He also guided me through the framed pictures on the wall—massive crowds of people protesting open-pit mining (a pitched battle Casa Pueblo helped win); images from their forest school where they do outdoor education; scenes from a protest in Washington, D.C., against a proposed gas pipeline through these mountains (another win). The community center was a strange hybrid of ecotourism lodge and revolutionary cell.

Settling into a wooden rocking chair, Massol-Deyá said that Maria had changed his sense of what's possible on the island. For years, he explained, he had pushed for the archipelago to get far more of its power from renewables. He had long warned of the risks associated with Puerto Rico's overwhelming dependence on imported fossil fuels and centralized power generation: One big storm, he had cautioned, could knock out the whole grid—especially after decades of laying off skilled electrical workers and letting maintenance lapse.

Now everyone whose homes went dark understood those risks, just as the people in Adjuntas could all look to a brightly lit Casa Pueblo and immediately grasp the advantages of solar energy, produced right where it is consumed. As Massol-Deyá put it: "Our quality of life was good before, because we were running with solar power. And after the hurricane, our quality of life is good as well. . . . This was an energy oasis for the community."

It's hard to imagine an energy system more vulnerable to climate change–amplified shocks than Puerto Rico's. The island gets an astonishing 98 percent of its electricity from fossil fuels. But since it has no domestic supply of oil, gas, or coal, all of these fuels are imported by ship. They are then transported to a handful of hulking power plants by truck and pipeline. Next, the electricity those plants generate is transmitted across huge distances through above-ground wires and an underwater cable that connects the island of Vieques to the main island. The whole behemoth is monstrously expensive, resulting in electricity prices that are nearly twice the U.S. average.

And just as environmentalists like Massol-Deyá had warned, Maria caused devastating ruptures within every tentacle of Puerto Rico's energy system: The Port of San Juan, which receives so much of the imported fuel, was thrown into crisis, and some 10,000 shipping containers full of much-needed supplies piled up on the docks, waiting to be delivered. Many truck drivers couldn't make it to the port, either because of obstructed roads, or because they were struggling to get their own families out of danger. With diesel in short supply across the island, some just couldn't find the fuel to drive. The lines at gas stations stretched out by the mile. Half of the island's stations were out of commission altogether. The mountain of supplies stuck at the port grew ever larger.

Meanwhile, the cable connecting Vieques was so damaged it had yet to be repaired six months later. And the power lines carrying electricity from the plants were down all over the archipelago. Literally nothing about the system worked.

This broad collapse, Massol-Deyá explained, was now helping him make the case for a sweeping and rapid shift to renewable energy. Because in a future that is sure to include more weather shocks, getting energy from sources that don't require sprawling transportation networks is just common sense. And Puerto Rico, though poor in fossil fuels, is drenched in sun, lashed by wind, and surrounded by waves.

Renewable energy is by no means immune to storm damage. At some Puerto Rican wind farms, turbine blades snapped off in Maria's high winds (seemingly because they were improperly positioned), just as some poorly secured solar panels took flight. This vulnerability is partly why Casa

Pueblo and many others emphasize the micro-grid model for renewables. Rather than relying on a few huge solar and wind farms, with power then carried over long and vulnerable transmission lines, smaller, community-based systems would generate power where it is consumed. If the larger grid sustains damage, these communities can simply disconnect from it and keep drawing from their micro-grids.

This decentralized model doesn't eliminate risk, but it would make the kind of total power outage that Puerto Ricans suffered for months—and which hundreds of thousands are suffering still—a thing of the past. Whoever's solar panels survive the next storm would, like Casa Pueblo, be up and running the next day. And "solar panels are easy to replace," Massol-Deyá pointed out—unlike power lines and pipelines.

In part to spread the gospel of renewables, in the weeks after the storm, Casa Pueblo handed out 14,000 solar lanterns—little square boxes that recharge when left outside during the day, providing a much-needed pool of light by night. More recently, the community center has managed to distribute a large shipment of full-sized solar-powered refrigerators, a game-changer for households in the interior that still don't have power.

Casa Pueblo has also kicked off #50ConSol, a campaign calling for 50 percent of Puerto Rico's power to come from the sun. They have been installing solar panels on dozens of homes and businesses in Adjuntas, including, most recently, a barbershop. "Now we have houses asking us for support," Massol-Deyá said—a marked shift from those days not so long ago when Casa Pueblo's solar panels looked like eco-luxury items. "We're going to do whatever is at reach to change that landscape and to tell the people of Puerto Rico that a different future is possible."

Several Puerto Ricans I spoke with casually referred to Maria as "our teacher." Because amid the storm's convulsions, people didn't just discover what didn't work (pretty much everything). They also learned very quickly about a few things that worked surprisingly well. Up in Adjuntas, it was solar power. Elsewhere, it was small organic farms that used traditional farming methods that were better able to stand up to the floods and wind. And in every case, deep community relationships, as well as strong ties to the Puerto Rican diaspora, successfully delivered lifesaving aid when the government failed and failed again.

Casa Pueblo was founded 38 years ago by Arturo's father, Alexis Massol-González, who was awarded the prestigious Goldman Prize for environmental leadership in 2002. Massol-González shares his son's belief that Maria has opened up a window of possibility, one that could yield a fundamental shift to a healthier and more democratic economy—not just for electricity, but also for food, water, and other necessities of life. "We are looking to transform the energy system. Our goal is to adopt a solar energy system and leave behind oil, natural gas, and carbon," he said, "which are highly polluting."

His message particularly resonates 45 miles to the southeast, in the coastal community of Jobos Bay, near Salinas. This is one of the areas coping with a slew of environmental toxins, much of it stemming from antiquated fossil fuel-burning power plants. As in Adjuntas, residents here have seized on the post-Maria electricity failures to advance solar power, through a project called Coquí Solar. Working with local academics, they have developed a plan that would not only produce enough energy to meet their needs, but would also keep the profits and jobs in the community as well. Nelson Santos Torres, one of Coquí Solar's organizers, told me they are insisting on solar skills training "so that community youth can participate in the installation," giving them a reason to stay on the island.

When I visited the area, Mónica Flores, a graduate student in environmental sciences at the University of Puerto Rico who has been working with communities on renewable energy projects, told me that truly democratic resource management is the island's best hope. People need to have a sense, she said, that "this is our energy. This is our water, and this is how we manage it because we believe in this process, and we respect our culture, our nature, everything that is supporting us."

Months into the rolling disaster set off by Maria, dozens of grassroots organizations are coming together to advance precisely this vision: a reimagined Puerto Rico run by its people in their interests. Like Casa Pueblo, in the myriad dysfunctions and injustices the storm so vividly exposed, they see opportunities to tackle the root causes that turned a weather disaster into a human catastrophe. Among them: the island's extreme dependence on imported fuel and food; the unpayable and possibly illegal debt that has been used to impose wave after wave of austerity that gravely weakened the island's defenses; and the 130-year-old colonial relationship with a U.S. government that has always discounted the lives of Puerto Rico's Black and Brown people.

If Maria is a teacher, this emerging movement argues, the storm's overarching lesson is that now is not the moment for reconstruction of what was, but rather for transformation into what could be. "Everything we consume comes from abroad and our profits are exported," said Massol-González, his hair now white after decades of struggle. It's a system that leaves debt and austerity behind, both of which made Puerto Rico exponentially more vulnerable to Maria's blows.

But, he said with a mischievous smile, "we look at crisis as an opportunity to change."

Massol-González and his allies know well that they are not alone in seeing opportunity in the post-Maria moment. There is also another, very different version of how Puerto Rico should be radically remade after the storm, and it is being aggressively advanced by Gov. Ricardo Rosselló in meetings with bankers, real estate developers, cryptocurrency traders, and, of course, the Financial Oversight and Management Board, an un-elected seven-member body that exerts ultimate control over Puerto Rico's economy.

For this powerful group, the lesson that Maria carried was not about the perils of economic dependency or austerity in times of climate disruption. The real problem, they argue, was the public ownership of Puerto Rico's infrastructure, which lacked the proper free-market incentives. Rather than transforming that infrastructure so that it truly serves the public interest, they argue for selling it off at fire-sale prices to private players.

This is just one part of a sweeping vision that sees Puerto Rico transforming itself into a "visitor economy," one with a radically downsized state and many fewer Puerto Ricans living on the island. In their place would be tens of thousands of "high-net-worth individuals" from Europe, Asia, and the U.S. mainland, lured to permanently relocate by a cornucopia of tax breaks and the promise of living a five-star resort lifestyle inside fully privatized enclaves, year-round.

In a sense, both are utopian projects—the vision of Puerto Rico in which the wealth of the island is carefully and democratically managed by its people, and the libertarian project some are calling "Puertopia" that is being conjured up in the ballrooms of luxury hotels in San Juan and New York City. One dream is grounded in a desire for people to exercise collective sovereignty over their land, energy, food, and water; the other in a desire for a small elite to secede from the reach of government altogether, liberated to accumulate unlimited private profit.

As I traveled throughout Puerto Rico, from sustainable farms and schools in the central mountain region, to the former U.S. Navy base on Vieques, to a legendary mutual aid center on the east coast, to former sugar plantations-turned-solar farms in the south, I found these very different visions of the future sprinting to advance their respective projects before the window of opportunity opened up by the storm begins to close.

At the core of this battle is a very simple question:

Who is Puerto Rico for? Is it for Puerto Ricans, or is it for outsiders? And after a collective trauma like Hurricane Maria, who has a right to decide?

INVASION OF THE PUERTOPIANS

Earlier this month, in San Juan's ornate Condado Vanderbilt Hotel, the dream of Puerto Rico as a for-profit utopia was on full display. From March 14 to 16, the hotel played host to Puerto Crypto, a three-day "immersive" pitch for blockchain and cryptocurrencies with a special focus on why Puerto Rico will "be the epicenter of this multitrillion-dollar market."

Among the speakers was Yaron Brook, chair of the Ayn Rand Institute, who presented on "How Deregulation and Blockchain Can Make Puerto Rico the Hong Kong of the Caribbean." Last year, Brook announced that he had personally relocated from California to Puerto Rico, where he claims he went from paying 55 percent of his income in taxes to less than 4 percent.

Elsewhere on the island, hundreds of thousands of Puerto Ricans were still living by flashlight, many were still dependent on FEMA for food aid, and the island's main mental health hotline was still over-whelmed with callers. But inside the sold-out Van-derbilt conference, there was little space for that kind of downer news. Instead, the 800 attendees—fresh from a choice between "sunrise yoga and medita-tion" and "morning surf"—heard from top officials like Department of Economic Development and Commerce Secretary Manuel Laboy Rivera about all the things Puerto Rico is doing to turn itself into the ultimate playground for newly minted crypto-currency millionaires and billionaires.

It's a pitch the Puerto Rican government has been making to the private jet set for a few years now, though until recently it was geared mainly to the financial sector, Silicon Valley, and others capa-ble of working wherever they can access data. The pitch goes like this: You don't have to relinquish your U.S. citizenship or even technically leave the United States to escape its tax laws, regulations, or the cold Wall Street winters. You just have to move your company's address to Puerto Rico and enjoy a stunningly low 4 percent corporate tax rate—a fraction of what corporations pay even after Don-ald Trump's recent tax cut. Any dividends paid by a Puerto Rico–based company to Puerto Rican res-idents are also tax-free, thanks to a law passed in 2012 called Act 20.

Conference attendees also learned that if they move their own residency to Puerto Rico, they will not only be able to surf every single morning, but also win vast personal tax advantages. Thanks to a clause in the federal tax code, U.S. citizens who move to Puerto Rico can avoid paying federal income tax on any income earned in Puerto Rico. And thanks to another local law, Act 22, they can also cash in on a slew of tax breaks and total tax waivers that includes paying zero capital gains tax and zero tax on interest and dividends sourced to Puerto Rico. And much more—all part of a desperate bid to attract capital to an island that is functionally bankrupt.

To quote billionaire hedge fund magnate John Paulson, owner of the hotel in which Puerto Crypto took place, "You can essentially minimize your taxes in a way that you can't do anywhere else in the world." (Or, as the tax dodger's website Premier Offshore put it: "All the other tax havens might as well just close down. . . . Puerto Rico just hit it out of the park . . . did the best set ever and dropped the mic.")

With just a 3 1/2-hour commute from New York City to San Juan (or less, depending on the private jet), all it takes to get in on this scheme is agreeing to spend 183 days of the year in Puerto Rico—in other words, winter. Puerto Rican residents, it's worth noting, are not only excluded from these programs, but they also pay very high local taxes.

Manuel Laboy used the conference to announce the creation of a new advisory council to attract blockchain businesses to the island. And he extolled the lifestyle bonuses that awaited attendees if they followed the self-described "Puertopians" who have already taken the plunge. As Laboy told The Intercept, for the 500 to 1,000 high-net-worth individuals who relocated since the tax holidays were introduced five years ago—many of them opting for gated communities with their own private schools—it's all about "living in a tropical island, with great people, with great weather, with great piña coladas." And why not? "You're gonna be, like, in this endless vacation in a tropical place, where you're actually working. That combination, I think, is very powerful."

The official slogan of this new Puerto Rico? "Paradise Performs." To underscore the point, conference attendees were invited to a "Cryptocurrency Honey Party," with pollen-themed drinks and snacks, and a chance to hang out with Ingrid Suarez, Miss Teen Panama 2013 and upcoming contestant on "Caribbean's Next Top Model."

Mining cryptocurrencies is one of the fastest growing sources of greenhouse gas emissions on the planet, with the industry's energy consumption rising by the week. Bitcoin alone currently consumes roughly the same amount of energy per year as Israel, according to the Bitcoin Energy Consumption Index. The city of Plattsburgh, New York, recently adopted a temporary ban on cryptocurrency mining after local electricity rates suddenly soared. Many of the crypto companies currently relocating to Puerto Rico would presumably do their currency mining elsewhere. Still, the idea of turning an island that cannot keep the lights on for its own people into "the epicenter of this multitrillion-dollar market" rooted in the most wasteful possible use of energy is a bizarre one and is raising mounting concerns of "crypto-colonialism."

In part to allay these fears, Puerto Crypto made a last-minute name change to the less imperial "Blockchain Unbound," though it didn't stick. Moreover, for some in the crypto crowd, the appeal of relocating to Puerto Rico goes well beyond Laboy's version of paradise. Post-Maria, with land selling for even cheaper, public assets being auctioned at fire-sale prices, and billions in federal disaster funds flowing to contractors, some distinctly more grandiose dreams for the island have begun to surface. Now rather than simply shopping for mansions in resort communities, the Puertopians are looking to buy a piece of land large enough to start their very own city—complete with airport, yacht port, and passports, all run on virtual currencies.

Some call it "Sol," others call it "Crypto Land," and it even seems to have its own religion: an unruly hodgepodge of Ayn Randian wealth supremacy, philanthrocapitalist noblesse oblige, Burning Man pseudo-spirituality, and half-remembered scenes from watching "Avatar" while high. Brock Pierce, the child actor turned crypto-entrepreneur who serves as the movement's de facto guru, is known for dropping New Age aphorisms like, "A billionaire is someone who has positively impacted the lives of a billion people." Out on a real estate expedition scouting locations for Crypto Land, he reportedly crawled into the "bosom" of a Ceiba tree, a magnificent species sacred in many indigenous cultures, and "kissed an old man's feet."

But make no mistake—the true religion here is tax avoidance. As one young crypto-trader recently told his YouTube audience, before moving to Puerto Rico in time to make the tax-filing deadline, "I had to actually look it up on the map." (He subsequently admitted to some "culture shock" upon learning that Puerto Ricans spoke Spanish, but instructed viewers thinking of following his lead to put a "Google translator app on your phone and you're good to go.")

The conviction that taxation is a form of theft is not a novel one among men who imagine themselves to be self-made. Still, there is something about rapidly becoming rich from money that you literally created—or "mined"—yourself that lends an especially large dose of self-righteousness to the decision to give nothing back. As Reeve Collins, a 42-year-old Puertopian, told the *New York Times*, "This is the first time in human history anyone other than kings or governments or gods can create their own money." So who is the government to take any of it from them?

As a breed, the Puertopians, in their flip-flops and surfer shorts, are a sort of slacker cousin to the Seasteaders, a movement of wealthy libertarians who have been plotting for years to escape the grip of government by starting their own city-states on artificial islands. Anybody who doesn't like being taxed or regulated will simply be able to, as the Seasteading manifesto states, "vote with your boat."

For those harboring these Randian secessionist fantasies, Puerto Rico is a much lighter lift. When it comes to taxing and regulating the wealthy, its current government has surrendered with un-matched enthusiasm. And there's no need to go to the trouble of building your own islands on elabo-rate floating platforms—as one Puerto Crypto ses-sion put it, Puerto Rico is poised to be transformed into a "crypto-island."

Sure, unlike the empty city-states Seasteaders fantasize about, real-world Puerto Rico is densely habited with living, breathing Puerto Ricans. But FEMA and the governor's office have been doing their best to take care of that too. Though there has been no reliable effort to track migration flows since Hurricane Maria, some 200,000 people have reportedly left the island, many of them with federal help.

This exodus was first presented as a temporary emergency measure, but it has since become apparent that the depopulation is intended to be permanent. The Puerto Rican governor's office predicts that over the next five years, the island's population will experience a "cumulative decline" of nearly 20 percent.

The Puertopians know all this has been hard on locals, but they insist that their presence will be a blessing for the devastated island. Brock Pierce argues (without offering any specifics), that cryptomoney is going to help finance Puerto Rican reconstruction and entrepreneurship, including in local agriculture and energy. The enormous brain drain currently flowing out of Puerto Rico, he says, is now being offset with a "brain gain," thanks to him and his tax-dodging friends. At a Puerto Rico investment conference, Pierce observed philosophically that "it's in these moments where we experience our greatest loss that we have our biggest opportunity to sort of restart and upgrade."

Gov. Rosselló himself seems to agree. In February 2018, he told a business audience in New York that Maria had created a "blank canvas" on which investors could paint their very own dream world.

AN ISLAND WEARY OF OUTSIDE EXPERIMENTS

The dream of the blank canvas, a safe place to test one's boldest ideas, has a long and bitter history in Puerto Rico. Throughout its long colonial history, the archipelago has continuously served as a living laboratory for prototypes that would later be exported around the globe. There were the notorious experiments in population control that, by the mid-1960s, resulted in the coercive sterilization of more than one-third of Puerto Rican women. Many dangerous drugs have been tested in Puerto Rico over the years, including a high-risk version of the birth control pill containing a dosage of hormones four times greater than the drugs that ultimately entered the U.S. market.

Vieques—more than two-thirds of which used to be a U.S. Navy facility where Marines practiced ground warfare and completed their gun training—was a testing ground for everything from Agent

Orange to depleted uranium to napalm. To this day, agribusiness giants like Monsanto and Syngenta use the southern coast of Puerto Rico as a sprawling testing ground for thousands of trials of genetically modified seeds, mostly corn and soy.

Many Puerto Rican economists also make a compelling case that the island invented the whole model of the special economic zone. In the '50s and '60s, well before the free-trade era swept the globe, U.S. manufacturers took advantage of Puerto Rico's low-wage workforce and special tax exemptions to relocate light manufacturing to the island, effectively road testing the model of offshored labor and maquiladora-style factories while still technically staying within U.S. borders.

The list could go on and on. The appeal of Puerto Rico for these experiments was a combination of the geographical control offered by an island and straight-up racism. Juan E. Rosario, a longtime community organizer and environmentalist who told me that his own mother was a Thalidomide test subject, put it like this: "It's an island, isolated, with a lot of nonvaluable people. Expendable people. For many years, we have been used as guinea pigs for U.S. experiments."

These experiments have left indelible scars on Puerto Rico's land and people. They are visible in the shells of factories that were abandoned when

U.S. manufacturers got access to even cheaper wages and laxer regulations in Mexico and then China after the North American Free Trade Agreement was signed and the World Trade Organization was created. The scars are etched too in the explosive materials, uncleared munitions, and diverse cocktail of military pollutants that will take decades to flush from Vieques's ecosystem, as well as in the small island's ongoing health crisis. And they are there in the swaths of land all over the archipelago that are so contaminated that the Environmental Protection Agency has classified 18 of them as Superfund sites, with all the local health impacts that shadow such toxicity.

The deepest scars may be even harder to see. Colonialism itself is a social experiment, a multilayered system of explicit and implicit controls designed to strip colonized peoples of their culture, confidence, and power. With tools ranging from the brute military and police aggression used to put down strikes and rebellions, to a law that once banned the Puerto Rican flag, to the dictates handed down today by the unelected fiscal control board, residents of these islands have been living under that web of controls for centuries.

On my first day on the island, at a meeting of trade union leaders at the University of Puerto Rico, Rosario spoke passionately about the psychological

impact of this unending experiment. He said that
at such a high-stakes moment—when so many outsiders are descending wielding their own plans and
their own big dreams—"we need to know where are
we heading. We need to know where is our ultimate goal. We need to know what paradise looks
like." And not the kind of paradise that "performs"
for currency traders with a surfing hobby, but that
actually works for the majority of Puerto Ricans.

The problem, he went on, is that "people in
Puerto Rico are very fearful of thinking about the
Big Thing. We are not supposed to be dreaming;
we are not supposed to be thinking about even gov-
erning ourselves. We don't have that tradition of
looking at the big picture." This, he said, is colonialism's most bitter legacy.

The belittling message at the core of the colonial
experiment has been reinforced in countless ways
by the official responses (and nonresponses) to Hurricane Maria. Time after humiliating time, Puerto
Ricans have been sent that familiar message about
their relative worth and ultimate disposability. And
nothing has done more to confirm this status than
the fact that no level of government has seen fit to
count the dead in any kind of credible way, as if lost
Puerto Rican lives are of so little consequence that
there is no need to document their mass extinguish-
ment. As of this writing, the official count of how

many people died as a result of Hurricane Maria remains at 64, though a thorough investigation by Puerto Rico's Center for Investigative Journalism and the *New York Times* put the real number at well over 1,000. Puerto Rico's governor has announced that an independent probe will re-examine the official numbers.

But there is a flipside to these painful revelations. Puerto Ricans now know, beyond any shadow of a doubt, that there is no government that has their interests at heart, not in the governor's mansion, not on the unelected fiscal control board (which many Puerto Ricans welcomed at first, convinced it would root out corruption), and certainly not in Washington, where the current president's idea of aid and comfort was to hurl paper towels into a crowd. That means that if there is to be a grand new experiment in Puerto Rico, one genuinely in the interest of its people, then Puerto Ricans themselves will have to be the ones to dream it up and fight for it—"from the bottom to the top," as Casa Pueblo founder Alexis Massol-González told me.

He is convinced that his people are up to the task. And ironically, this is in part thanks to Maria. Precisely because the official response to the hurricane has been so devoid of urgency, Puerto Ricans on the island and in the diaspora have been forced to organize themselves on a stunning scale. Casa Pueblo is just one example among many. With next to no resources, communities have set up massive communal kitchens, raised large sums of money, coordinated and distributed supplies, cleared streets, and rebuilt schools. In some communities, they have even gotten the electricity reconnected with the help of retired electrical workers.

They shouldn't have had to do all this. Puerto Ricans pay taxes—the IRS collects some $3.5 billion from the island annually—to help fund FEMA and the military, which are supposed to protect U.S. citizens during states of emergency. But one result of being forced to save themselves is that many communities have discovered a depth of strength and capacity they did not know they possessed.

Now this confidence is rapidly spilling over into the political arena and with it, an appetite among a growing number of Puerto Rican groups and individuals to do precisely what Juan E. Rosario said has been so difficult in the past: come up with their own big ideas, their own dreams of an island paradise that performs for them.

"WELCOME TO MAGIC LAND"

Those were the words that greeted me at a bustling public school and organic farm carved into the hillside in Puerto Rico's spectacular central mountain region, a place known for its towering waterfalls, crystal natural pools, and electric green peaks.

After driving for an hour and a half through communities still badly battered by the hurricane, the scene did feel strangely enchanted. There were smiling children harvesting a crop of beans and wandering through stands of sunflowers. There were young men and women sawing lumber and busily erecting several new structures, stopping periodically to share ideas about how to get the farm working to maximum potential. And in a region where many are still relying on inadequate government food aid, there were older women preparing mountains of vegetables and fish for a sumptuous communal meal.

The mood was so upbeat and the efficiency so

undeniable that I had a feeling similar to the one I had at Casa Pueblo—as if I had stepped through a portal to that parallel Puerto Rico, a place where both the ecological and economic lessons of Hurricane Maria were being powerfully heeded.

"We do agro-ecological farming," Dalma Cartagena told me, pointing to the rows of spinach, kale, cilantro, and much more. "Kids from third grade to eighth grade do this work, this beautiful work."

Cartagena—a trained agronomist with braided gray curls and a yogic smile—is most passionate about how farming has helped her students overcome the trauma of a storm that was so ferocious, it felt as if the natural world had turned against them. Running her fingers through a stand of medicinal flowers, she said, "After Maria, we encourage the students to touch the plants and let the plants touch them because that's a way of healing the pain and anger."

When students watch plants grow that they planted from seeds, it's a reminder that despite all of the damage inflicted by the storm, "You are part of something that is always protecting you." The apparent rupture between themselves and the land begins to heal.

Eighteen years ago, Cartagena took charge of this farm in the municipality of Orocovis as part of the Puerto Rico Education Department's embattled "agriculture education program." Connected by a short pathway to a large local middle school, Escuela Segunda Unidad Botijas I, students spend part of each day on the farm, listening to Cartagena explain everything from the nitrogen cycle to composting. Dressed in neat school uniforms complemented with mud-caked rubber boots, they also learn the practical skills of "agro-ecology," a term referring to a combination of traditional farming methods that promotes resilience and protects biodiversity, a rejection of pesticides and other toxins, and a commitment to rebuilding social relationships between farmers and local communities.

Each grade tends to their own crops from seed to harvest. Some of what they grow is served in the school cafeteria, some is sold at market, and most goes home with the students.

Concentrating through heavy, black-framed glasses as she shelled a pile of beans, 13-year-old Brítany Berríos Torres explained, "My mom can make them, or she can give them to my grandmother so she can stop worrying about 'What am I going to cook my daughters?'" With so much need on the island, doing this work, Torres said, "I feel as if we are throwing a rope to humanity."

All of this makes this public school's farm a relative anomaly in Puerto Rico. As a legacy of the

slave plantation economy first established under
Spanish rule, much of the island's agriculture is in-
dustrial scale, with many crops grown for export or
testing purposes. Roughly 85 percent of the food
Puerto Ricans actually eat is imported.

With her unique school, which the government has tried to shut down several times, Cartagena is determined to prove that this dependency
on outsiders is not only unnecessary, but a kind of
folly. By using farming techniques and carefully
preserved seed varieties adapted to the region, she
is convinced that Puerto Ricans can feed themselves with healthy food grown in their own fertile
soil—as long as there is sufficient land available for
a new and existing generation of farmers with the
knowledge to do the work.

This lesson of self-sufficiency took on very
practical urgency after Hurricane Maria. Just as the
upheaval revealed the perils of Puerto Rico's import-
addicted and highly centralized energy system, it
also unmasked the extraordinary vulnerability of
its food supply. All over the island, industrial-scale
farms growing mono-crops of banana, plantains,
papaya, coffee, and corn looked like they had been
flattened with a scythe. According to Puerto Rico's
Department of Agriculture, more than 80 percent
of the island's crops were completely wiped out in
the storm, a $2 billion blow to the economy.

"A lot of conventional farmers right now are starving, even though they have [an] amazing amount of land," Katia Avilés, an environmental geographer and agro-ecological farming advocate, told me. "They didn't have anything to harvest because they had followed the Department of Agriculture's instructions" and literally bet the farm on a single, vulnerable cash crop.

Food imports, meanwhile, were in no better shape. The Port of San Juan was in chaos, with shipping containers filled with desperately needed food and fuel sitting unopened. For weeks, the shelves at many supermarkets were virtually empty. Remote areas like Orocovis fared the worst: stranded because of blocked roads and insufficient fuel, it took over a week or more for food aid to arrive. And when it came, it was often shockingly inadequate: military-style rations and FEMA's now notorious boxes filled with Skittles, processed meats, and Cheez-It crackers.

On Cartagena's small farm, however, there was nutritious food to share. The storm had knocked down the greenhouse and her outdoor classroom, and the wind had claimed the bananas. But many of the crops the students had planted were fine: the tomatillos, the root vegetables—pretty much everything that grows low to the earth or underneath it.

"We never closed the farm. We stayed here

working," Cartagena said, "cleaning up and doing the compost, the way we could." Within days, students began crossing the mountains by foot to help out, carrying food home to their families. They planted flowers to try to lure back the bees.

There was other help too. On the day I visited, the land was crowded with about 30 farmers who had traveled from across the United States, Central America, Canada, and Puerto Rico to help Cartagena and her students rebuild and replant. The visitors were part of a wave of international "brigades" that had been going from farm to farm rebuilding chicken coops, greenhouses, and other outdoor structures, as well as replanting crops, an ambitious effort organized by Puerto Rico's Organización Boricuá de Agricultura Ecológica, the U.S.-based Climate Justice Alliance, and the global network of peasants and small farmers, Via Campesina.

Jesús Vázquez, an environmental justice advocate, food sovereignty activist and local coordinator of the brigades, told me that Cartagena's experience was not unique. In the days after Maria, farmers and community members helped one another across the island. And those rare estates that still used traditional methods—including planting a diversity of crops and using trees and grasses with long roots to prevent landslides and erosion—had some of the only fresh food on the island.

Yucca, taro, sweet potato, yam, and several other root vegetables are nutrient-rich staples of the Puerto Rican diet, and because they grow underground, where the high winds couldn't touch them, most were almost entirely protected from storm damage. "Some farmers were harvesting food a day after the hurricane," Vázquez recalled. Within a few weeks, they had hundreds of pounds of food to sell or distribute in their communities.

Avilés, Vázquez, and Cartagena all work with Organización Boricuá, a network of farmers who use these traditional Puerto Rican methods, passing them down through the generations, "campesino to campesino," as Avilés put it. But after decades of U.S. government policy that equated campesino life with underdevelopment and set Puerto Rico up as a captive market for U.S. imports, all that remains, Avilés said, are "islands" of these agro-ecological farms scattered through the archipelago's three inhabited islands.

For 28 years, Organización Boricuá has been connecting those farming islands to one another, advocating for their interests and publicly making the case that agro-ecology should form the basis for Puerto Rico's food system, capable of providing "adequate, affordable, nutritious, and culturally appropriate food" for the entire population, Vázquez explained. The group has also been warning about the dangers of chokepoints in Puerto Rico's highly centralized system, with almost all of its food imports shipping out of a single port in Jacksonville, Florida (which itself was slammed by Hurricane Irma last September), and roughly 90 percent of the food arriving at one entry point: the Port of San Juan. "We've always been saying within our movement that that's a problem because of climate change," Vázquez told me. After all, if something happens to the port, "then we'll be doomed."

Given the strength of the corporate agricultural lobbies they were up against, getting these kinds of messages through to the public has been an uphill battle. Their opponents painted them as backward relics, while imports and fast food were modernization incarnate. But Maria, which was powerful enough to rearrange local geology, has changed the political topography as well.

Overnight, everyone could see just how dangerous it was for this fertile island to have lost control over its agricultural system, along with so much else. "We didn't have food, we didn't have water, we didn't have electricity, we didn't have anything," Avilés recalled. But in communities that still had traditional farms, people could also see that agro-ecology was not some quaint relic of the past, but a crucial tool for surviving a rocky future.

Now Organización Boricuá is joining with

many others who have been constructing their own "islands" of self-sufficiency—not just farms, but also solar-powered oases like Casa Pueblo, as well as mutual aid centers and groups of educators and economists with plans for how Puerto Ricans can confront international capital and remake their economy and public institutions. Together, this network of grassroots Puerto Rican movements is laying out a plan for a new Puerto Rico, one in which residents play a greater role in shaping their own destinies than they have at any time since the island was colonized by Spain in 1493. "It's just one fight," Katia Avilés said, "which is, how do we make sure that we have a just recovery and that for the future, we're not going to fall as hard as we did this time?"

And there will be a next time. I spoke with Elizabeth Yeampierre, executive director of UPROSE, Brooklyn's oldest Latino community-based organization, who was also in Puerto Rico as part of the climate justice brigades. She was preoccupied with the knowledge that hurricane season would begin again in just a few months. "It's impossible to talk about what happened in Puerto Rico without talking about climate change," which, by causing oceans to warm and sea levels to rise, is sure to bring more record-breaking storms. "It would be foolish for us to think that this is the

last storm, that there aren't going to be other re-
curring extreme weather events."

She also said that Puerto Ricans—by drawing on long-protected indigenous knowledge about
what seeds and tree species can survive extreme
events, as well as the kind of energy and sturdy social structures that can withstand these shocks—are
creating a model not just for the island, but for the
world. A way to "start really thinking about how
you prepare for the fact that climate change is here."

But if Puerto Rico's people's movements are
going to have a chance to provide this kind of
global leadership, they will need to move fast. Because they aren't the only ones with radical plans
about how the island should transform after Maria.

SHOCK-AFTER-SHOCK-AFTER-SHOCK DOCTRINE

The day before I walked through that portal in Orocovis, Gov. Ricardo Rosselló delivered a televised address from behind his desk, flanked by the flags of the United States and Puerto Rico. "While overcoming adversity, we also find great opportunities to build a new Puerto Rico," he announced. The first step was to be the immediate privatization of the Puerto Rico Electric Power Authority, known as PREPA, one of the largest public power providers in the United States and, despite its billions of dollars in debt, the one that brings in the most revenue.

"We will sell PREPA's assets to the companies that will transform the power generation system into a modern, efficient, and less costly system for our people," Rosselló said.

It turned out to be the first shot in a machine-gun loaded with such announcements. Two days later, the slick, TV-friendly young governor unveiled his long-awaited "fiscal plan," which included closing more than 300 schools and shutting down more than two-thirds of the island government's executive-branch entities, going from a total of 115 to just 35. As Kate Aronoff reported for The Intercept, this "amounts to a deconstruction of the island's administrative state" (so it's no surprise that Rosselló has many admirers in Trump's Washington).

A week after that, the governor went on television again and unveiled a plan to crack open the education system to privately run charter schools and private school vouchers—moves Puerto Rico's teachers and parents have successfully resisted several times before.

This deliberate exploitation of states of emergency to push through a radical pro-corporate agenda is a phenomenon I have called the "shock doctrine." And it is playing out in Puerto Rico in the most naked form seen since New Orleans's public school system and much of its low-income housing were dismantled in the immediate aftermath of Hurricane Katrina, while the city was still largely empty of its residents. And Puerto Rico's education secretary, the former management consultant Julia Keleher, makes no secret of where she is drawing inspiration from. One month after Maria, she tweeted that New Orleans should be a "point of reference," and "we should not underestimate the damage or the opportunity to create new, better schools."

Central to a shock doctrine strategy is speed—pushing a flurry of radical changes through so quickly it's virtually impossible to keep up. So, for instance, while most of the meager media attention has focused on Rosselló's privatization plans, an equally significant attack on regulations and independent oversight—laid out in his fiscal plan—has gone largely under the radar.

And the process is far from complete. There is a great deal of talk about more privatizations to come: highways, bridges, ports, ferries, water systems, national parks, and other conservation areas. Manuel Laboy, Puerto Rico's secretary of economic development and commerce, told The Intercept that electricity is just the beginning. "We do expect that similar things will happen in other infrastructure sectors. It could be full privatization; it could be a true P3 [public-private partnerships] model."

Despite the radical nature of these plans, the response from Puerto Rican society has been somewhat muted. No large-scale protests greeted the first wave of Rosselló's rapid-fire announcements. No strikes in response to his plans to radically contract the state and roll back pensions. No uprisings against the Puertopians flooding into the island to build their libertarian dream state.

Yet Puerto Rico has a deep history of popular resistance and some very radical trade unions. So what is going on? The first thing to understand is that Puerto Ricans are not experiencing one extreme dose of the shock doctrine, but two or even three of them, all layered on top of one another—a new and terrifying hybridization of the strategy that makes it particularly challenging to resist.

Many Puerto Ricans told me that the latest chapter in this story really begins in 2006, when the tax breaks that had been used to attract U.S. manufacturers to the island were allowed to expire, prompting a devastating wave of capital flight (and demonstrating just how precarious it is to build a development policy based on tax giveaways). This was such a deep shock to the island's economy that in May 2006, much of the government, including all the public schools, was temporarily shut down. That was the first punch. The second came when the global financial system melted down less than two years later, dramatically deepening a crisis already well underway.

Broke and desperate, the Puerto Rican government turned to borrowing, in part by using its special tax status to issue municipal bonds that were exempt from city, state, and federal taxes. It also purchased high-risk capital appreciation bonds, which will eventually rack up interest rates ranging from 785 to 1,000 percent. Thanks in large part to these kinds of predatory financial instruments, borrowed under conditions that many experts argue were illegal under the Puerto Rican Constitution, the island's debt exploded. According to data compiled by lawyer Armando Pintado, debt-service payments, including interest and other profits paid to the banking industry, increased fivefold between 2001 and 2014, with a particularly marked spike in 2008. Yet another shock to the island's economy.

And so, in that all-too-familiar shock doctrine story, an atmosphere of crisis was exploited to force severe austerity on a desperate people. In 2009, Puerto Rico's governor passed a law declaring an economic "state of emergency" and used it to lay off more than 17,000 public sector workers and strip negotiated benefits and raises from many more—this at a time when unemployment was already 15 percent. As has been the case everywhere these policies have been imposed in recent years—from the U.K. to Greece—it didn't bring the island back to growth and health. It pushed it deeper into joblessness, recession, and bankruptcy.

It was in this context that in 2016, Congress took the drastic measure of passing the PROMESA law that put Puerto Rico's finances under the control of a newly created Financial Oversight and Management Board, a seven-person body appointed by the U.S. president, six of whom appear not to live on the island. The board, which is essentially charged with overseeing the liquidation of Puerto Rico's assets to maximize debt repayments and approving all major economic decisions, is known in Puerto Rico as "La Junta." For many, the name is a commentary on the fact that the board represents a kind of financial coup d'état: Puerto Ricans—unable to vote for president or Congress but forced to live under U.S. laws—already lacked basic democratic rights. By giving the fiscal board the power to reject decisions made by Puerto Rico's elected territorial representatives, they were now losing the weak rights they had won, marking a return to unmasked colonial rule.

Unsurprisingly, the fiscal control board promptly placed Puerto Rico on an even more wrenching austerity diet. It demanded deep cuts to pensions and public services, including health care, as well as a laundry list of privatizations. The school system was particularly hard-hit in this period. Between 2010 and 2017, roughly 340 public schools were shut down; arts and physical education programs were virtually eliminated in many elementary schools; and the board announced plans to slash the University of Puerto Rico's budget in half.

Yarimar Bonilla, a Rutgers University associate professor who had been conducting a major research project on Puerto Rico's debt crisis before Maria hit, told me there is no way to understand the post-Maria shock doctrine strategy without recognizing that Puerto Ricans "were already in a state of shock and severe economic policies were already being applied here. The government had already been whittled down and people's expectations for the government had already been very much whittled down." By early 2017, she pointed out, parts of San Juan looked very much like they had been hit by a hurricane—windows were broken, buildings were boarded up. But it wasn't high winds that did it; it was debt and austerity.

Perhaps the most relevant part of this story, however, is that by 2017, Puerto Ricans were resisting this shock doctrine strategy with organization and militancy. There had been resistance at earlier stages, including a general strike in 2009. But in the months before Maria struck, Puerto Rico saw some of the strongest and most unified opposition in the island's history.

A popular movement calling for an independent audit of the debt was quickly gaining ground, spurred by the conviction that if its causes were closely examined, as much as 60 percent of the more than $70 billion Puerto Rico supposedly owes would be found to have been accumulated in violation of the island's constitution and is therefore illegal. And if a large part of the debt is illegal, not only would it need to be erased, the fiscal control board would need to be dismantled, and debt could no longer be used as a cudgel with which to impose austerity and further weaken democracy. According to Eva Prados, spokesperson for the Citizens Front for the Audit of the Debt, in the year before Hurricane Maria, 150,000 Puerto Ricans added their names to a call to audit the debt, and thousands participated in vigils calling for "light and truth."

And then there was the mounting revolt against austerity. Last spring, students at the University of Puerto Rico's 11 campuses staged a historic strike that lasted more than two months, protesting plans to raise tuition while their school's budget was being slashed, as well as the broader austerity agenda. A faculty group launched a major lawsuit against the fiscal control board alleging that the deep cuts to the university were an illegal attack on an essential service. Then, on May 1, 2017, many of Puerto Rico's labor and social movements converged into one angry cry, when roughly 100,000 people took to the streets to demand an end to austerity and an audit of the debt—by some estimates, the second-largest protest in Puerto Rico's history.

It was clear that this movement had authorities worried. After several banks were vandalized, the state launched an intense crackdown against the key organizations involved in the May 1 anti-austerity mobilization, threatening them with costly lawsuits and jailing several activists.

In this atmosphere of heated resistance, with many calling for Rosselló's resignation, several of the more draconian plans seemed to stall. The cuts to the university were in question, as were some of the bigger-ticket privatizations. The secretary of education, meanwhile, had been forced to scale back the number of planned public school closures. Not every battle was won, but it was clear that there would be no all-out shock doctrine–style makeover of Puerto Rico without a fight.

Then came Maria, and all those same rejected policies came roaring back with Category 5 ferocity.

DESPERATION, DISTRACTION, DESPAIR, AND DISAPPEARANCE

The jury is still out as to whether this latest attempt at the shock-after-shock doctrine approach will actually work. If it does, it will not be because Puerto Ricans suddenly overwhelmingly approve of these policies. It will be because the tremendous impact of the storm has disassembled life for millions of people, making the reconstitution of the pre-storm, anti-austerity coalition a herculean challenge.

It's helpful to break the extreme state of shock that is being exploited into four categories: desperation, distraction, despair, and disappearance.

Desperation because the relief and reconstruction efforts have been so sluggish, so inept, and so apparently corrupt that they have understandably instilled a sense in many that nothing could be worse than the status quo. This is particularly true for electricity. Even among those that have had their power restored, many are experiencing regular blackouts. They are also hearing daily threats from their governor that the whole island could wind up back in the dark again at any point because PREPA is so broke that it can't pay the bills; in some parts of the island, water is being rationed for similar reasons. It's circumstances like these that make the prospect of privatization more palatable. With the status quo so untenable, anything at all can seem like an improvement.

Related to this is distraction: Daily life in Puerto Rico remains an immense struggle. There are repairs to be done to damaged homes, and byzantine, time-devouring bureaucracies to navigate to help pay for them. For those who still don't have electricity or water, there are the interminable lineups required to receive aid. Many workplaces still remain closed, making paying the bills yet another huge logistical hurdle, if it's possible at all. Add all this together and for many Puerto Ricans, the mechanics of survival can take up every waking hour—a state of distraction not very conducive to political engagement.

For many, the burdens of survival have been so onerous, and future prospects seemingly so bleak, that a deep despair has set in—indeed it is reaching epidemic proportions. Callers making credible threats to take their own lives overwhelmed the island's 24-hour mental health hotline in the months after the hurricane. According to a government report, more than 3,000 people who called the line between November 2017 and January 2018 reported having already attempted suicide—a 246 percent increase over the previous year.

For Yarimar Bonilla, these figures represent not just the impacts of Maria, devastating as they have been, but rather the cumulative effects of many compounding blows. "Puerto Ricans had already undergone a huge amount of trauma due to the colonial relationship to the United States," most recently during the debt crisis. Then came the storm, which literally ripped the lid off the agony that so many households had been quietly enduring. With cameras poking into homes that had their roofs torn apart, Puerto Ricans found themselves looking into one another's lives, and they saw not just storm damage, but also punishing poverty, untreated illness, and social isolation. As Bonilla put it, "There's a real sorrow here in a place that used to be known for its joy."

Today, she says, there may not be rioting in the streets, but that should not be confused with consent. The apparent passivity is at least partly the result of so much pain being directed inward.

The same desperate circumstances have forced

hundreds of thousands of Puerto Ricans to make the wrenching decision to simply disappear from the island. They vanish daily onto planes headed for Florida and New York and elsewhere in the mainland United States. Many of them have had the direct help of FEMA, which built what the agency called an "air bridge," airlifting people off the island and boarding others onto cruise ships. Once on the mainland, they were provided with funds to stay in hotels (supports were set to expire on March 20).

Bonilla says this approach was a political choice—much as it was a choice to fly and bus the residents of New Orleans to distant states after Hurricane Katrina, often offering no way to return, a process that permanently changed the demographics of the city. "Instead of helping people here, providing shelters here, bringing more generator power to the places that need them, getting the electric system up and running, they're encouraging people to leave instead."

There are several reasons why evacuation may have been heavily favored by Washington and the governor's office. The disappearance of so many people in such a short time, Bonilla explained, "operates as a political escape valve, so right now you don't have people protesting in the streets because a lot of the people who are really desperate for medical care or who had real needs where they couldn't live without electricity have just left."

The exodus also conveniently helps create the "blank canvas" that the governor has bragged about to would-be investors. Elizabeth Yeampierre helped welcome and support many of her fellow Puerto Ricans when they arrived in the United States. But when I spoke with her on the island, she said that her "biggest fear" is that the evacuation will be a prelude to a massive land grab. "What they want is our land, and they just don't want our people in it."

Many Puerto Ricans I spoke with are similarly convinced that there is more than incompetence behind the various ways they are being pushed to the limits of endurance.

As has been extensively reported since the storm hit, the relief and reconstruction efforts have been a nonstop procession of almost impossibly disastrous decisions. A key contract to supply 30 million meals went to an Atlanta company with a record of failure and a staff of one (only 50,000 meals were delivered before the contract was canceled). Desperately needed relief supplies sat for weeks in storage, both in San Juan and Florida, where some became rat-infested. Materials key to rebuilding the electrical grid also sat in warehouses for unknown reasons. Whitefish Energy, a Montana-based firm with ties to Interior Secretary Ryan Zinke, had just two full-time staff when it landed a $300 million contract to help rebuild the electricity grid (the contract has since been canceled).

Then there were the common-sense measures that were simply passed over. As many pointed out, the Trump administration could have swiftly sent in the USNS *Comfort*, a massive floating hospital, to ease the strain on failing health care facilities. Instead, the ship was sent in late, sat nearly empty for weeks, and then was ordered withdrawn in November, with power still out on half of the island. Similarly, instead of relying on two-bit contractors like Whitefish, or notorious profiteers like Fluor, which has cashed in on disasters from post-invasion Iraq to post-Katrina New Orleans, PREPA could have requested that other state electrical utilities send workers to Puerto Rico and help with the rebuilding—its right as a member of the American Public Power Association. But it waited more than a month before putting in the request.

Each one of these decisions, even when they were ultimately reversed, set recovery efforts back further. Is this all a masterful conspiracy to make sure Puerto Ricans are too desperate, distracted, and despairing to resist Wall Street's bitter economic medicine? I don't believe it's anything that coordinated. Much of this is simply what happens when you bleed the public sphere for decades, laying off competent workers and neglecting basic maintenance. Run-of-the-mill corruption and cronyism are no doubt at work as well.

But it's also true that many governments have deployed a starve-then-sell strategy when it comes to public services: cut health care/transit/education to the bone until people are so disillusioned and desperate that they are willing to try anything, including selling off those services altogether. And if Rosselló and the Trump administration have seemed remarkably unconcerned about the nonstop relief and reconstruction screw-ups, the attitude may be at least partly informed by an understanding that the worse things get, the stronger the case for privatization becomes.

Mónica Flores, the University of Puerto Rico graduate student researching renewable energy, said the whole experience has been like watching a car wreck in slow motion. Like so many others, Flores said it felt impossible to take on these systemic issues when you have lost your home, when you are living out of your car, when you are going to friends' houses to shower. "You're trying not to fall apart . . . and people are immobilized because they're scared, because they're lost, because they're just trying to survive."

Many Puerto Ricans point out that the promises of lower prices and greater efficiency that would flow from privatizing basic services are contradicted by their own experiences. Private telephone companies have provided poor service in many parts of the archipelago, and a water and sewage system sale in the '90s proved so economically and environmentally disastrous, it had to be reversed less than a decade later. Many fear this experience will be repeated—that if PREPA is privatized, the Puerto Rican government will lose an important source of revenue, while getting stiffed with the utility's multibillion-dollar debt. They also fear that electricity rates will stay high, and that poor and remote regions where people are less able to pay could well lose access to the grid altogether.

Even so, the governor's pitch has proved persuasive for some because privatization is not presented as one possible solution to a dire humanitarian crisis, but as the only one. As Casa Pueblo and Coquí Solar are attempting to show, this is far from the truth. There are other energy models—implemented successfully in countries like Denmark and Germany—that would greatly improve Puerto Rico's broken and dirty state-run utility, while keeping power and wealth in the hands of Puerto Ricans. But advancing such democratic models requires the political participation of a population that has a lot of other things on its plate right now.

There is reason to hope, however, that a post-Maria shock-resistance may be starting to take root. Mercedes Martínez, the indomitable head of the Federation of Puerto Rican Teachers, has spent the months since the storm crisscrossing the island, warning parents and educators that the plan to radically downsize and privatize the school system relies upon their fatigue and trauma.

While visiting a still-closed school in Humacao, in the eastern region, she told a local teacher that the government "knows we're made of flesh and bones—they know that human beings get worn out and discouraged." But, she insisted, if people understand that it is a strategy, they can defeat it.

"Our job is to motivate people to know that it's possible to resist things as long as we believe in ourselves." This was more than a pep talk: In the few months after Maria, the secretary of education attempted to keep dozens of schools from reopening, claiming they were unsafe. The teachers feared it was a prelude to closing the schools for good.

Again and again, parents and teachers—who had, in many cases, repaired the buildings themselves—successfully fought to protect their local schools. "They occupied the schools, reopened them without permission; parents blocked the streets," Martínez recalled. As a result, more than 25 schools were reopened that the government had tried to close for good after the storm.

That's why Martínez is convinced that no

matter what is written in the governor's fiscal plan and no matter what privatization laws have been introduced, it is still possible for Puerto Ricans to successfully resist the shock doctrine. Especially if the pre-storm coalitions rebuild and expand.

On March 19, 2018, teachers across Puerto Rico held a one-day walkout to protest the plans to shrink and privatize the island's school system, the first major political demonstration since Maria. And talk of a full-blown strike was growing louder.

I asked Martínez if her members feared taking action that would disrupt the lives of families that have already been through so much. She was un-equivocal. "Absolutely not. Our feeling is, how can the government add more pain to children's lives by shutting down their schools, taking away their teachers, and setting up a privatized system that favors those who already have the most?"

THE ISLANDS OF SOVEREIGNTY CONVERGE

On my last day in Puerto Rico, we climbed another mountain and stepped through yet another
portal. I was traveling with Sofía Gallisá Muriente,
a Puerto Rican artist I had first met in the Rocka-
ways in the aftermath of Superstorm Sandy, where
she had been part of the grassroots relief effort
known as Occupy Sandy.

We'd been scaling treacherously narrow roads
on the east coast of the island, taking various
wrong turns because many signs were still down,
looking for the community center in the village of
Mariana. Finally, we asked a man on the side of
the road for directions. "You mean the breadfruit
festival? It's right up there."

We found ourselves in a clearing with hundreds
of people from across the archipelago, gathered on
folding chairs under a large, white tent. From up
here, looking down the valley to the sea, we could

see precisely where Maria first made landfall.

As the roadside confusion suggested, this was indeed the site of an annual festival that celebrates a large, starchy, and nutritious fruit, one that attracts hundreds of people for food and music to this village in the municipality of Humacao every year. But after the area was left without food aid for 10 days, only to get boxes filled with Skittles, the festival's kitchen facilities were harnessed for a different use: Women who usually do the cooking for the festival came together, pooled whatever food they could find, and made hot, healthy meals for about 400 people a day. Day after day. Week after week. Month after month. They are doing it still.

Renamed the Proyecto de Apoyo Mutuo Mariana (the Mutual Aid Project of Mariana), the center has become a symbol of the miracles Puerto Ricans have been quietly pulling off while their governments fail them. In addition to the communal kitchen, which brought the neighborhood together around meals, the project started organizing brigades to go out and clear debris. Next, they set up programing for kids, since the schools were still closed.

Christine Nieves, a dynamic thinker who left a post at Florida State University's business school to move back to the island a year before the storm, is one of the forces behind this project. She and her partner, musician Luis Rodríguez Sánchez, used their contacts off-island to turn the community center into a functioning hub, with solar panels and backup batteries, a Wi-Fi network, water filters, and rainwater cisterns.

Since Mariana still doesn't have power or water, the mutual aid center at the top of the mountain has become yet another energy oasis, the only place to plug in electronics and medical equipment. The next stage for the project, Nieves told me, is to extend solar power to other buildings in the community in a micro-grid.

The biggest challenge, she said, has been helping people to see that they don't need to wait for others to solve problems—everyone has something they can contribute now. They might not have food or water, she went on, but people know how to do things. "You know electricity? Actually, we have a problem that you can help us with. You know plumbing?" That's a skill they can put to use, too.

This process of discovering the latent potential in the community has been like "opening your eyes and all of a sudden seeing 'Oh wait, we're humans and there's other ways of relating to each other [now that] the system has stopped,'" Nieves said.

I came here to see this remarkable project, but also because on this day, Proyecto de Apoyo Mutuo Mariana was hosting several hundred organizers and intellectuals from across Puerto Rico, as well as a couple dozen visitors from the United States and Central America. Convened by PAReS, a collective of University of Puerto Rico faculty members involved in the anti-austerity struggle, the meeting had been billed as a gathering of organizations and movements "against disaster capitalism and for other worlds."

It was the first time movements had gathered across such a broad spectrum since Maria changed everything. And many observed that it was the first chance they had had in months to step back, take stock, and strategize. "We organized the gathering in this post-Maria moment to be able to look at each other, talk, and see if we could come together at this crossroads to create a different future," Mariolga Reyes Cruz, a PAReS collective member and a contingent faculty at the Río Piedras campus, told me.

People gathered here from all the parallel worlds I visited during my time in Puerto Rico, all the islands hidden away in these islands. I saw farmers from Organización Boricuá, determined to show that given the right supports, they can feed their own people without relying on imports; solar warriors from Casa Pueblo and Coquí Solar, who have seized the moment to push a rapid transition to locally controlled renewables; teachers who have organized their communities to keep their schools open. And tired and muddy members of the solidarity brigades that had come to help rebuild.

Key leaders from last year's surge of anti-austerity activism were here too—organizers from the student strike, the lawyers and economists calling for an audit of Puerto Rico's debt, trade union leaders and academics who had been researching alternatives for Puerto Rico's economy for a long time.

After a brief welcome, the organizers assigned discussion themes before breaking everyone up into smaller groups spread out in clusters on the mountaintop. Snippets of conversations floated up from these working groups: "We need reinvention not reconstruction" . . . "We can't just defend the public as if it's inherently good" . . . "We need a moratorium on any attempt to fast-track private schools" . . . "A just recovery means not just responding to the disaster, but to the underlying *causes* of the disaster."

Surveying the scene, Christine Nieves told me that that it felt like "a dream come true that we didn't know we had." She added, "I think we're going to look back to this moment"—when such a wide diversity of groups, most of whom did not know each other before the storm, all came together "in this beautiful, open space, wondering how do we create an alternative and building toward an alternative"—and realize that this was the moment when things shifted from despair to possibility.

As the groups reconvened to share their findings,

it was possible to detect an emerging synthesis—or at least, a better understanding of how the various fronts on which Puerto Ricans are fighting fit into a larger whole. The debt must be audited because by calling its legality into question, the case to abolish the anti-democratic fiscal control board, and all of its endless demands for "structural reforms," grows stronger. And that's crucial because Puerto Ricans can't exercise their sovereignty if they are subject to the whims of a body they had no hand in electing.

For generations, the struggle for national sovereignty has defined politics in Puerto Rico: Who favors independence from Washington? Who wants to become the 51st state, with full democratic rights? Who defends the status quo? So it seems significant that as discussions unfolded in Mariana, a broader definition of freedom emerged. I heard talk of "multiple sovereignties"—food sovereignty, liberated from dependence on imports and agribusiness giants; energy sovereignty, liberated from fossil fuels and controlled by communities. And perhaps housing, water, and education sovereignty as well.

What also seemed to be growing was an understanding that this decentralized model is even more important in the context of climate change, where islands like this one will be buffeted by many more extreme events capable of knocking out centralized systems of all kinds, from communication networks to electricity grids to agricultural supply chains.

The day ended with shared food cooked in the community kitchen: rice and beans, mashed taro, stewed cod, homemade rum flavored with every fruit in the island's rainbow. Next came live trovador music and dancing until long after dark. As volunteers helped clean up the kitchen, an elderly neighbor arrived to quietly plug in his oxygen machine and have a chat with friends.

Watching this mass meeting segue seamlessly into a party, I was reminded of Yarimar Bonilla's observation that amid Puerto Rico's epidemic of despair, "the people who seem to be doing the best are those who are helping others, those who are involved in community efforts." That was certainly true here. And it was true, too, of the young people I met in Orocovis, bursting with pride about how they were able to bring food home to their families.

It makes sense that helping would have this healing effect. To live through a profound trauma like Maria is to know the most extreme form of powerlessness. For what felt like an eternity, families were unable to reach one another to find out if their loved ones were alive or dead; parents were unable to protect their children from harm. It stands to reason that the best cure for helplessness is . . . helping, being a participant, rather than a spectator, in the recovery of your home, community, and land.

This is why the shock doctrine, as a political strategy, is more than just cynical and opportunist—"it's cruel," as Mónica Flores said to me through tears. By forcing people to watch as their shared resources are sold out from under them, unable to stop it because they are too busy trying to survive, the disaster capitalists who have descended on Puerto Rico are reinforcing the most traumatizing part of the disaster they are there to exploit: the experience of helplessness.

RACE AGAINST TIME

Earlier in the day in Mariana, one speaker had described the challenge they faced as a kind of race between "the speed of movements and the speed of capital."

Capital is fast. Unencumbered by democratic norms, the governor and the fiscal control board can whip up their plan to radically downsize and auction off the territory in a matter of weeks—even faster, in fact, because their plans were fully developed during the debt crisis. All they had to do was dust them off and repackage them as hurricane relief, then release their fiats. Hedge fund managers and crypto-traders can similarly decide to relocate and build their "Puertopia" on a whim, with no one to consult but their accountants and lawyers.

Which is why the "Paradise Performs" version of Puerto Rico is moving along at such a rapid clip. For instance, I interviewed Keith St. Clair, a fast-talking Brit who moved to the island to take advantage of the tax breaks and began investing in hotels. He told me that he had met with the governor shortly after Maria. "And I said, 'I'm gonna double down, I'm gonna triple down, I'm gonna quadruple down, because I believe in Puerto Rico.'" Looking out at the virtually empty Isla Verde Beach in front of one of his San Juan hotels ("a 90 percent tax-exempt property"), he predicted, "This could be Miami, South Beach. . . . That's what we are trying to create."

The grassroots groups here in Mariana are entirely unconvinced that becoming a fly-in bedroom community for tax-dodging plutocrats represents any kind of serious economic development strategy. And they fear that if this post-disaster gold rush is allowed to continue unchecked, it will foreclose on the very different versions of paradise they are daring to imagine for their island.

Land is scarce in Puerto Rico, especially prime farmland. If it all gets snapped up for more office towers, malls, hotels, golf courses, and mansions, there will only be scraps left for sustainable farms and renewable energy projects. And if infrastructure spending is poured into toll-road highways, high-priced ferries, and airports, there similarly won't be anything left for public transit and a local food system. Moreover, if energy privatization goes ahead, it could become prohibitively costly for local communities to pursue the solar and wind micro-grid model. After all, private utility companies from Nevada to Florida have successfully pressured their state governments to put up roadblocks to renewables, since a market in which your customers are also your competitors (able to generate their own power and sell it back to the grid) is distinctly less profitable. Rosselló's fiscal plan already floats the idea of a new tax that would penalize communities that set up their own renewable micro-grids.

All of these are fateful choices. Manuel Laboy, Puerto Rico's secretary of economic development, said that the decisions made in this window "are going to basically set the principles and the conditions for the next 50 years."

The trouble is that movements, unlike capital, tend to move slowly. This is particularly true of movements that exist to deepen democracy and allow ordinary people to define their goals and grab the reins of history.

It's a very good thing, then, that Puerto Ricans are not beginning to build this movement for self-determination from scratch. Indeed, they have been preparing for this moment for generations, from the height of the independence struggle to the successful battle to kick the U.S. Navy out of Vieques, to the anti-austerity and anti-debt coalition that peaked in the months before Maria.

And many Puerto Ricans have also been building their future world in miniature, on those islands of sovereignty hidden throughout the territory. Now, in Mariana, those islands have found each other, forming their own parallel political archipelago.

Elizabeth Yeampierre, who attended the Mariana summit, believes that despite all the devastation being visited on Puerto Rico, her people have the fortitude for the battles ahead. "I see a level of resistance and support that I didn't imagine was going to be possible," she said. "And it reminds me that these are the descendants of colonization and slavery, and they are strong."

In the weeks after I left the island, the 60 groups represented in Mariana solidified into a political bloc that they named JunteGente (the People Together) and have had meetings all over the archipelago. Inspired by different models around the world, they have begun drafting a people's platform, one that will unite their various causes into a common vision for a radically transformed Puerto Rico. It is grounded in an unabashed insistence that despite centuries of attacks on their sovereignty, the Puerto Rican people are the only ones who have the right to dream up their collective future.

And so, six months after Maria revealed so much that didn't work and a few important things that did,

Puerto Rico finds itself locked in a battle of utopias. The Puertopians dream of a radical withdrawal from society into their privatized enclaves. The groups that gathered in Mariana dream of a society with far deeper commitments and engagement—with each other, within communities, and with the natural systems whose health is a prerequisite for any kind of safe future. In a very real sense, it's a battle between sovereignty for the many versus secession for the few.

For now, these diametrically opposed versions of utopia are advancing in their own parallel worlds, at their own speeds—one on the back of shocks, the other in spite of them. But both are gaining power fast, and in the high-stakes months and years to come, collision is inevitable.

ACKNOWLEDGMENTS

The Intercept team: Betsy Reed, Roger Hodge (story editor), Charlotte Greensit, Sharon Riley (researcher and fact-checker), Lauren Feeney, Andrea Jones, Philipp Hubert.

The Haymarket team: Julie Fain, Brian Baughan, Teresa Córdova Rodríguez (Spanish translation), Rachel Cohen (cover and interior design), Jim Plank, and Anthony Arnove, who made this whole project possible.

As well as: Jackie Joiner, Avi Lewis, Angela Adrar, Katia Avilés, Federico Cintrón Moscoso, Gustavo García López, Ana Elisa Pérez, Mariolga Reyes Cruz, Juan Carlos Rivera Ramos, Jesús Vázquez, Elizabeth Yeampierre, Ruth Santiago, Bernat Tort Ortiz, Carmen Yulín Cruz, José La Luz, Sofía Gallisá Muriente, Eva Prados, Cristian Carretero, Eduardo Mariota, Ana Tijoux, the Climate Justice Alliance, UPROSE, Casa Pueblo,

Organización Boricuá de Agricultura Ecológica, and The Leap.

Deepest thanks go to the engaged intellectuals at PAReS, for inviting me to Puerto Rico to help amplify these stories.

The Intercept_

After NSA whistleblower Edward Snowden came forward with revelations of mass surveillance in 2013, journalists Glenn Greenwald, Laura Poitras, and Jeremy Scahill decided to launch a new media organization dedicated to the kind of reporting those disclosures required: fearless, adversarial journalism. They called it The Intercept.

Today, The Intercept (theintercept.com) is an award-winning news organization that covers national security, politics, civil liberties, the environment, international affairs, technology, criminal justice, the media, and more. Led by Editor-in-Chief Betsy Reed, its reporters have the editorial freedom to hold powerful institutions accountable—and the support they need to pursue investigations that expose corruption and injustice.

Regular contributors include Mehdi Hasan, Naomi Klein, Shaun King, Sharon Lerner, James Risen, Liliana Segura, and co-founding editors Glenn Greenwald and Jeremy Scahill. EBay founder and philanthropist Pierre Omidyar provided the funding to launch The Intercept and continues to support it through First Look Media Works, a nonprofit organization.

HAYMARKET BOOKS

Haymarket Books is a radical, independent, nonprofit book publisher based in Chicago.

Our mission is to publish books that contribute to struggles for social and economic justice. We strive to make our books a vibrant and organic part of social movements and the education and development of a critical, engaged, international left.

We take inspiration and courage from our namesakes, the Haymarket martyrs, who gave their lives fighting for a better world. Their 1886 struggle for the eight-hour day—which gave us May Day, the international workers' holiday—reminds workers around the world that ordinary people can organize and struggle for their own liberation. These struggles continue today across the globe—struggles against oppression, exploitation, poverty, and war.

Since our founding in 2001, Haymarket Books has published more than five hundred titles. Radically independent, we seek to drive a wedge into the risk-averse world of corporate book publishing. Our authors include Noam Chomsky, Arundhati Roy, Rebecca Solnit, Angela Y. Davis, Howard Zinn, Amy Goodman, Wallace Shawn, Mike Davis, Winona LaDuke, Ilan Pappé, Richard Wolff, Dave Zirin, Keeanga-Yamahtta Taylor, Nick Turse, Dahr Jamail, David Barsamian, Elizabeth Laird, Amira Hass, Mark Steel, Avi Lewis, Naomi Klein, and Neil Davidson. We are also the trade publishers of the acclaimed Historical Materialism Book Series and of Dispatch Books.

ALSO AVAILABLE FROM HAYMARKET BOOKS

No Is Not Enough: Resisting Trump's Shock Politics
and Winning the World We Need
Naomi Klein

Sin Patrón: Stories from Argentina's Worker-Run Factories
lavaca collective, foreword by Naomi Klein and Avi Lewis

Modern Colonization by Medical Intervention:
U.S. Medicine in Puerto Rico
Nicole Trujillo-Pagán

Freedom Is a Constant Struggle: Ferguson, Palestine,
and the Foundations of a Movement
Angela Y. Davis

Things That Make White People Uncomfortable
Michael Bennett and Dave Zirin
Foreword by Martellus Bennett

How We Get Free: Black Feminism
and the Combahee River Collective
Edited by Keeanga-Yamahtta Taylor

Photo by Kourosh Keshiri

ABOUT THE AUTHOR

Naomi Klein is an award-winning journalist, syndicated columnist, documentary filmmaker, and author of the international bestsellers *No Logo*, *The Shock Doctrine: The Rise of Disaster Capitalism*, *This Changes Everything: Capitalism vs. the Climate*, and *No Is Not Enough*. She is Senior Correspondent for The Intercept and contributor to *The Nation* magazine. She is also a Puffin Foundation Writing Fellow at The Nation Institute and her writing appears widely in such publications as the *New York Times*, *Le Monde*, and the *Guardian*. Klein is a member of the board of directors for climate-action group 350.org and one of the organizers behind Canada's Leap Manifesto (theleap.org), a blueprint for a rapid and justice-based transition off fossil fuels. In November 2016 she was awarded Australia's prestigious Sydney Peace Prize. Her books have been translated into more than 30 languages.